GLOBAL WARMING:
Manmade or Man Made-up?

Joseph L. Swick

GLOBAL WARMING:
Manmade or Man Made-up?

Joseph L. Swick

First Edition

GLOBAL WARMING: MANMADE OR MAN MADE-UP?
by **Joseph L. Swick**

Published by
Swickstar Enterprizes
Rt. 1 Box 214
Philippi, WV 26416

Edition ISBN: 978-1-44950-451-9
First edition.
Printed in the United States of America.

Contents

FOREWORD
By Dr. G.L. Vincent, Ph.D.

When approached by Mr. Swick to comment on Global Warming as a Foreword to his book on the subject, a lot of images raced through my head: I thought about penguins in Antarctica whose breeding habits were disrupted because their ice bridges had melted and they were becoming lost. I thought about the canals of Venice being flooded and those historical buildings along its banks slowly going under water (and that was just the tip of the melting iceberg.) Where were we headed and what would I say about the subject?

Global warming is ambiguous and daunting. Personally, I am not really sure there is any easy solution to this problem.

I do know that I am not 100% sold on the idea that it is entirely created by humans, though we may have contributed to it. There are a variety of factors causing our world to warm. I think the public needs to be aware of the multiple factors that contribute to this problem.

As someone who has known Joseph Swick personally for almost a decade, I can say that he is passionate about the subject at hand. He wants the public to be informed, and to be enlightened that there is more than one reason for this phenomenon.

One should keep an open mind on this matter. Don't place too much stock in the news media. Think for yourself and make an intelligent *and informed* decision.

This book is an odyssey of different possible causes for global warming and a great collection to educate you on the subject.

Dr. Gary L. Vincent is an author of numerous books, is an international recording artist, and a concerned citizen for issues affecting our world. His latest book, **Surviving The Swine Flu**, is available on amazon.com. For more information, visit www.SurvivingTheSwineFlu.com.

CHAPTER 1
Pollution

Is the pollution in the air causing Global Warming? No one with common sense likes pollution. I've heard people complain about pollution and I don't like pollution. I don't like smelling rubbish burning in the neighbor's yard, being near an incinerator, or being near an industrial smoke stack, but we still have the need to ask ourselves the question: Is this contributing to Global Warming?

Nantong (city), China, setting under a cloud of smog.[1]

This has never been proven; this is only a theory. As a matter of fact, many scientists believe that Global Warming being

---

[1]   http://www.publicdomainpictures.net (photo by Peter Griffin)

entirely manmade is a myth. I share this belief. If you doubt that this is the case, then read on. My arguments will change your mind. If, on the other hand, we share a similar belief that there are many diverse reasons for Global Warming, my arguments will confirm your views.

My first example is the Chernobyl nuclear disaster which occurred in Russia in 1986. What effect did that have on the world's climate?

According to reports filed by the Soviet scientists investigating the biological and radiological affects of the Chernobyl disaster, the radioactive fallout levels within a 10 mile plus radius of the facility were extremely high. As a result, more than 4 square miles of forest habitat was subsequently destroyed. The ecological impact to the areas wildlife cannot be measured. This was a true ecological disaster.

Chernobyl – 20 years later[2]

High levels of radioactive isotopes were released into the atmosphere. But did this contribute to Global Warming?

The evidence indicates that the primary damage to the environment was the further depletion of the atmosphere's ozone levels. The long term effects are still being studied. We can only hope that lessons have been learned from this tragedy and that in the future similar events can be avoided.

One of the largest sources of airborne pollutants is automotive emissions. Anyone that has ever driven behind a large diesel truck can attest to this fact. However, some small levels of air pollutants are unavoidable if we are to travel and conduct commerce. The main problem is that we do not take full advantage of existing technologies that can reduce emissions. In this case, hydrogen fuel cells.

Hydrogen fuel cell technology has been around for quite some time. I've conducted research on this technology, and I believe it works and would be a practical solution. The byproduct of hydrogen fuel cells is water vapor.

So, if such a viable alternative to internal combustion engines exists, why are hydrogen fuel cells not widely used?

Answer: Common sense and big $.

We, as Americans, must use common sense and demand

change. Only through change will these cleaner technologies be implemented.

But do these emissions contribute to Global Warming?

Current evidence is inconclusive. Individuals on both sides of this debate are passionate. I believe that the truth lies somewhere in the middle. The key is for each of us to conduct our own research so that we can make intelligent informed decisions. Do not be swayed by politicians and media moguls that simply use existing information outlets as a platform to promote ideals that support their business interests. Decide for yourself!!

More research is needed before a definitive answer can be obtained. The hundreds of millions of dollars that are being spent each year to install anti-pollution devices on automobiles and factories should be spent to fund this research and develop the new technologies that would eliminate the problem instead of increasing production costs to our industrial center that costs this country tens of thousands of lost jobs and hundreds of millions of dollars in revenues. This would help rejuvenate the economy when it is desperately needed.

Another area that has received a great deal of media attention is agricultural pollution. New "studies" indicate that farms are producing toxic fumes and gases. Is the pollution from farming causing Global Warming? Humans have been involved in agriculture for thousands of years. So we are to believe that the agricultural industry during the last 30 years has damaged the ozone layer. This is simply preposterous.

So do politicians that espouse a Global Warming agenda have the best interests of America at heart, or are they simply profiting from the man made up idea?

Pollution is a serious problem and needs to be controlled. It can never be completely eliminated. Do I believe that we should return to the rampant pollution caused during the last two centuries of industrial progress? No. I think we should keep the controls on the emissions that are already in place. But we should not go to the extreme. Extreme measures cause extreme results. That is why our industrial center is in a state of decline. Industry cannot compete when excessive antipollution expenses cripple their bottom line.

Countries such as China have very few restrictions and regulations concerning pollution standards. Remember that their industrial emissions are adding to the global pollution problem. America is doing more, in my opinion, than just about any other country in the world to curtail pollution related problems. But where is the happy medium?

The antipollution Global Warming "industry" makes tens of millions of dollars each year. But is the threat real?

That's a question that you can answer yourself just by looking at our history. Geologists have clearly demonstrated that the Earth has undergone many warming and cooling epochs in its long history. So examine the real data and make your own decision. Is Global warming man made or man-made up?

Smokestacks from a wartime production plant,
World War II[3]

Raw sewage and industrial waste flows into the U.S. from Mexico as the New River passes from Mexicali, Baja California to Calexico, California.[4]

---

4   Calexico New River Committee (CNRC)

Soviet-era rocket blasting through the atmosphere.

CHAPTER 2
The Space Agencies

The second topic concerning Global Warming that I chose to discuss is the possibility that space program launches may be a contributing factor.

On October 4, 1957 the space age began with the launch of the first satellite (Sputnik-1). The Earth's ozone layer has deteriorated the most during the last 30 year span. Is there a correlation? Possibly. However, many other factors need to be considered before an educated conclusion can be reached.

One often cited possibility is the use of high and low sulfur coal. Close examination reveals that this is unlikely. During the last 150 years, the primary energy source used in the world shifted from coal burning industries and heating units to more energy efficient and environmentally safe alternatives. Thus it is unlikely that coal use is a major contributing factor to the "Global Warming" phenomenon.

Conversely, there are two probable contributing factors that did not exist in the distant past. The first is the advent of the space age. Damage to the ozone layer may be the result of the increased frequency of launches into space. As our reliance on satellites continues to grow, this problem may continue at an accelerated rate. The advent of space programs by additional nations, such as China, is also a likely contributing factor. Common sense dictates that the launch of materials through the ozone layer may indeed be causing damage.

Several years ago I viewed a televised interview with a NASA engineer. The question was raised if there was a possibility that space shuttle launches was a contributing factor in ozone depletion.

The engineer acknowledged that the space shuttle's solid rocket boosters did have an adverse impact on the ozone layer but he wouldn't state unequivocally that space travel was the major cause. He would not make a statement that could potentially jeopardize his career, but he did confirm the possibility.

Another unproven possibility is that nuclear power plants or explosions may be a candidate. Scientists espouse the merits of nuclear power. It is clean, renewable, and cost effective. But if one simply looks for plausible causes that did not exist in the distant past, nuclear power may be a possibility. Radioactive airborne isotopes from early nuclear bomb tests may also play a role. As stated previously, this is an unproven theory, but the time frames involved make an intriguing case.

You must conduct your own research and draw your own conclusions. Often, politicians use the threat of Global Warming as a platform to promote their own self-motivated political agendas. This is the perfect topic to choose as an ideological platform: It can't be proven either way. This is the primary reason that I am writing this book. Conduct your own research and use your best judgment when faced with a topic that you know little or nothing about. Use common sense to evaluate issues that could have far reaching impact and that could even change your life. Leave nothing to chance, be 100 percent certain.

So get motivated and find your own answers. Don't let popular media and politicians influence your decisions. When you allow another to do your thinking and simply agree with their conclusions, you find yourself controlled without realizing that anything has occurred.

CHAPTER 3
Dinosaurs

As I stated in an earlier chapter, the Earth has undergone drastic changes in temperature during its long history. There have been both warming and cooling trends. Keeping this in mind, I would like to ask any politician that has ever used Global Warming as an environmental platform to promote special interests one question: What caused the heating and cooling periods in the Earth's distant past? For example, what caused the environment to warm during the age of the Thunder Lizards, the dinosaurs? Where were the cars, buses, trains, factories, and other industries that caused the Global Warming? Isn't the answer obvious? What opinion polls were taken and what scientific studies were used to study the phenomenon? Answer: Natural climatic change that happens on a regular basis to planet Earth.

We have no control when the power of nature is unleashed. We have absolutely no control over natural weather conditions such as tornados, hurricanes, floods, blizzards or lightning. We simply have to accept that these events are inevitable and that there is very little that we can do to prevent them. We simply try to predict the patterns and give advance warnings so that people can escape out of harm's way. I contend that Global Warming is similar. Natural patterns of heating and cooling cycles have been impacting the earth since its creation. We simply are now entering a warming period. We are as likely to control the total effects of Global Warming as we are to control a hurricane. Our only controllable response is to reduce pollution emissions and hope for the best.

Let us now examine a cooling trend, otherwise known as an Ice Age. Thousands of years ago great mammals such as Woolly Mammoths reigned supreme. At that time, mean temperatures were much cooler and great glacial ice sheets extended well into the present United States from Canada. Eventually, weather patterns changed, the temperature rose, and the great ice sheets retreated. It can actually be argued that we are currently still experiencing the same great warming trend that ended the last Ice Age. Where were the machines and fires of industry that brought the Ice Age to an end? My contention is that we are over reacting to the natural evolution of climatic change on the Earth.

We simply do not have all of the facts. Our current understanding of the forces that govern weather patterns is growing but still is in its infancy.

We only have reliable weather data for the last few centuries of recorded history. A few centuries constitutes less than a millionth of 1 percent of the total life span of the Earth. We simply know very little about climate trends that could take thousands or millions of years to manifest. When you are listening to the raging debate concerning Global Warming keep these facts in mind and discover if you truly believe everything that you are being told. Conduct your own research and draw your own conclusions.

Then decide if Global Warming is manmade or man made up.

CHAPTER 4
Do the Believers Believe?

Do the proponents of Global Warming truly believe that it exists? Or is it merely a means to an end: the improvement of a profit margin. Think about this question for a moment. Do individuals that circle the global in private jets that guzzle fuel and spew toxic emissions really care about Global Warming?

If they truly believe in their own agenda, why not lead by

example. Cut their own energy consumption and do their part to save the planet. Is that happening? NO.

So are these individuals completely irresponsible, or merely uncaring when confronted with their own proposed environmental policies. This only makes common "cents". But they no longer have any common "cents"; it was spent to lease the private jet. Someone living in a ten million dollar mansion and driving a Rolls Royce is not going to tell me to drive an electric compact vehicle.

I don't know about you, but I have a difficult time dealing with hypocrites that do not "practice what they preach." I believe that you take a stand behind your beliefs, but not so far that you cannot recognize your own handiwork. I was once told by a used car salesperson that he stood behind everything that he sold. Far, far, behind it. Do you see a correlation? Use the common "cents" that you have been saving and draw your own conclusions.

Implementing environmental standards that cause increased

production costs of goods and services is not saving the planet. It merely shifts the industrial base when the factories close here and reopen in Asia

Using this reasoning, one would be under the impression that the United States existed under a magic bubble and would be impervious to air pollutants on the other side of the world. The environmental restrictions that we are adopting will only have a long term viable impact if ALL industrial nations agree to the same guidelines. Realistically, this will never happen. The production void created by the closure of industrial facilities in the United States is filled by an overseas concern. Based on the population growth curve in the world, demand for products is not decreasing and our economy continues to suffer while foreign interests thrive. Hence, the sum total impact of American environmental policy is to harm our own economy, enhance foreign businesses, and do little if anything to curtail the emission of greenhouse gases. So, is Global Warming man made or man made up?

By hikingArtist.com

CHAPTER 5
Profiting From Global Warming

If the political establishment is promoting the idea of Global Warming in order to earn additional profits, how are they doing it? By going GREEN. How often are you confronted with companies and products that are sold as an environmentally safe alternative? I would imagine nearly every day. Going Green has become big business. Saying an item is all natural or organic allows it to be marketed at a much higher price point under the guise of higher production costs. In this way, Global Warming actually becomes a sales pitch. How many people do you know that jump on the bandwagon for new fads because it is fashionable at the time? Going green is currently in fashion, so many companies are using this to their own advantage. Am I saying that we should try to make environmentally harmful products? No. I simply mean that we should not be trying to make unfair profits from a topic that is morally ambiguous. Of course we should create products, if possible, that do little or no harm to the ecosystem.

Everyone likes to see an image of a pristine, unspoiled, natural vista. Therefore, companies are in a win-win situation. By stating that a product is Green or organic, a false impression is created that implies that an item is better than its competitors when its actual performance may be equal or worse than a competitor's brand.

Another fashionable trend that is creating unbelievable

profits is bottled water. Ad agencies have created the false impression that bottled water is somehow better than water taken from the tap, when in actuality many bottled waters are filled from the tap. This is advertising genius at its finest.

An additional profitable enterprise is to conduct speaking tours about Global Warming that produce enormous profits from appearance fees for a topic that may not actually exist.

Many companies having been producing cleaning products for a hundred years or more, yet there is no definitive proof that their manufacturing techniques and products have contributed to Global Warming. Stop listening to the rhetoric, conduct your own research, and form an educated opinion.

So is Global Warming man made or man made-up?

Solar particles interact with Earth's magnetosphere[5]

5   http://sec.gsfc.nasa.gov/popscise.jpg

CHAPTER 6
Sunspots

Another naturally occurring phenomenon that provides compelling evidence for a role in triggering global warming and cooling epochs is sun spots. First, let me provide some basic information about sunspots.

According to Wikipedia, "A sun spot is an area on the sun's surface that is marked by intense magnetic activity, which inhabits forming areas of reduced surface temperature. They can be visible without the aid of a telescope.

"Although they are at temperatures of roughly 4,000–4,500 K, the contrast with the surrounding material at about 5,800 K leaves them clearly visible as dark spots, as the intensity of a heated black body (closely approximated by the photosphere) is a function of T (temperature) to the fourth power. If a sunspot were isolated from the surrounding photosphere it would be brighter than an electric arc.

"A minimum in the eleven-year sunspot cycle may have

occurred in December 2008, but the current lack of activity may push the minimum into 2009. While the reverse polarity sunspot observed on 4 January 2008 may represent the start of Cycle 24, only a few sunspots have yet been seen in this cycle. A new sunspot cycle occurs when the average number of sunspots of the new cycle's magnetic polarity outnumbers that of the old cycle's polarity. Forecasts in 2006 predicted Cycle 24 to start between late 2007 and early 2008; when this failed to manifest, new estimates suggested a delay until 2009. As of mid-June 2009, however, the expected increase in solar activity had not yet begun.

"Sunspots, being the manifestation of intense magnetic activity, host secondary phenomena such as coronal loops and reconnection events. Most solar flares and coronal mass ejections originate in magnetically active regions around visible sunspot groupings. Similar phenomena indirectly observed on stars are commonly called star spots and both light and dark spots have been measured."[6]

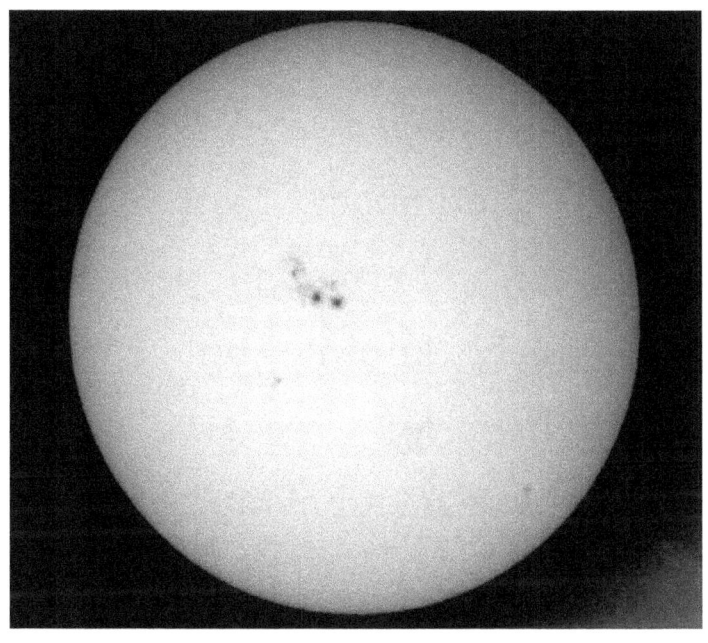

The effects of sunspots on the thermodynamic properties of the Earth's atmosphere have been proven. A mini ice age developed

---

6   http://en.wikipedia.org/wiki/Sunspot

in the last half of the 17th century that corresponded to unusual sunspot activity. It is widely believed that the sunspot frequency played a contributing factor. The impact of sunspots of the Earth's thermal properties is not widely known, but careful comprehensive research will uncover the relationship.

Sunspots were first reported by Chinese astronomers as early as 28 B. C., and also by Copernicus and Galileo. There seems to be an actual correlation between the predictable sunspot cycles and mean temperature changes. Thus, another potential natural trigger for heating cycles exists.

Leading economists have indicated that sunspot activity was linked to poor crop harvests in the later portion of the 19th century. This was caused by changes to the weather patterns brought about by unusual thermal currents within the jet stream. Geologic data also indicates that sunspot cycles directly impact the Earth's mean temperature. Therefore, how can anyone categorically state that greenhouse gas emissions are causing Global Warming when there are countless naturally occurring events that play a crucial role? So, is Global Warming man made or man made up?

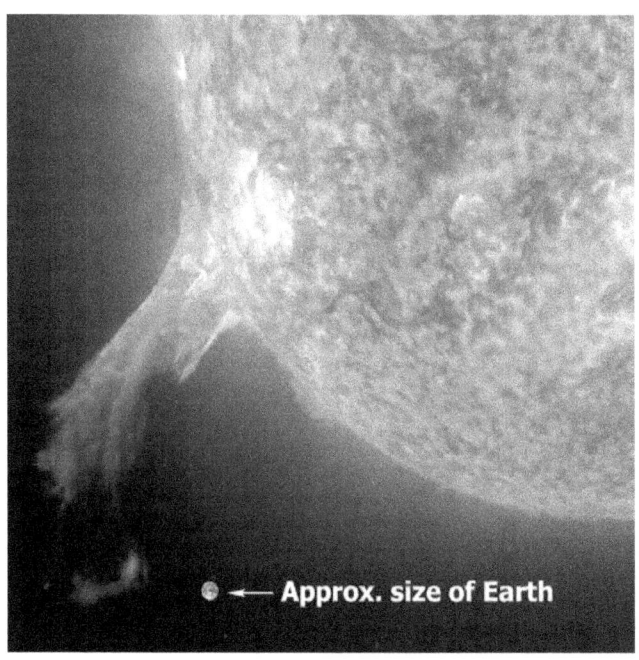

One of the sun's smaller flares in the last few years.[7]

Ash plume of Pinatubo during 1991 eruption.[8]

8   http://vulcan.wr.usgs.gov/Volcanoes/Philippines/Pinatubo/images.html

CHAPTER 7
Volcanoes

Volcanoes are another natural phenomenon that some scientists believe contribute to Global Warming.

The simplest explanation for volcanism is that it serves as a pressure release valve for the Earth's core. As temperatures and pressures rise within the Earth, weak fissures in the Earth's mantle will eventually give way and create a fault vent that will ultimately become what we call a volcano. Large scale eruptions of volcanoes and other sites of volcanism (such as Yellowstone) are responsible for causing millions of tons of ash and greenhouse gases to be released into the atmosphere. Volcanism is a contributing factor to the cycle of Global Warming and cooling.

According to Wikipedia:

"A volcano is an opening, or rupture, in a planet's surface or crust, which allows hot, molten rock, ash, and gases to escape from below the surface. Volcanic activity involving the extrusion of rock tends to form mountains or features like mountains over a period of time. The word volcano is derived from Italian vulcano, after Vulcan, the Roman god of fire.

"Volcanoes are generally found where tectonic plates are diverging or converging. A mid-oceanic ridge, for example the Mid-Atlantic Ridge, has examples of volcanoes caused by divergent tectonic plates pulling apart; the Pacific Ring of Fire has examples of volcanoes caused by convergent tectonic plates coming together. By contrast, volcanoes are usually not created where two tectonic plates slide past one another. Volcanoes can also form where there is stretching and thinning of the Earth's crust (called "non-hotspot intraplate volcanism"), such as in the African Rift Valley, the Wells Gray-Clearwater volcanic field and the Rio Grande Rift in North America and the European Rhine Graben with its Eifel volcanoes.

"Volcanoes can be caused by mantle plumes. These so-called hotspots, for example at Hawaii, can occur far from plate boundaries. Hotspot volcanoes are also found elsewhere in the solar system, especially on rocky planets and moons."[9]

There are many different types of volcanoes in the world and to some extent, they jettison gas and steam each day. This is naturally created air pollution that cannot be controlled. Marine volcanoes also play a role. Subsurface vents exist in all major oceans of the Earth. The unequal heating caused by these deep seas vents aids in the

[9]   http://en.wikipedia.org/wiki/Volcano

creation of ocean currents that have a direct bearing on weather patterns across the globe.

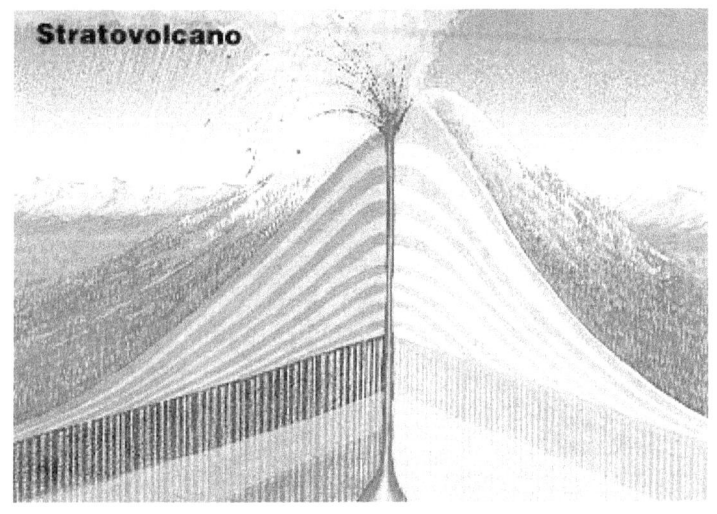

Stratovolcano[10]

This is but another example of a natural event that impacts the Earth's global temperature cycle. Volcanoes certainly cannot

---

[10] United States Geological Survey

be controlled and research continues concerning their impact on global weather events. So is global warming man made or man made-up?

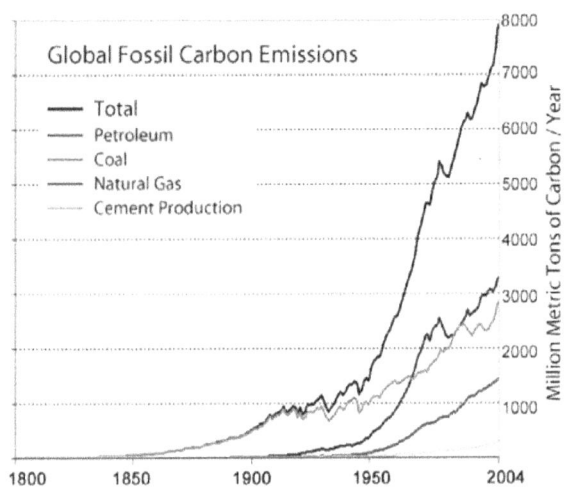

Global annual fossil fuel carbon dioxide emissions through year 2004, in million metric tons of carbon, as reported by the Carbon Dioxide Information Analysis Center[11]

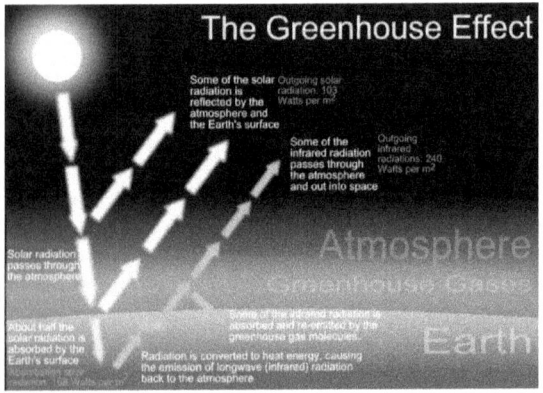

This diagram shows how the greenhouse effect works.[12]

11   http://en.wikipedia.org/wiki/Greenhouse_gas

12   http://en.wikipedia.org/wiki/File:The_green_house_effect.svg

CHAPTER 8
Greenhouse Gas

"Greenhouse gases are gases in an atmosphere that absorb and emit radiation within the thermal infrared range. This process is the fundamental cause of the greenhouse effect. Common greenhouse gases in the Earth's atmosphere include water vapor, carbon dioxide, methane, nitrous oxide, and ozone. In our solar system, the atmospheres of Venus, Mars and Titan also contain gases that cause greenhouse effects. Greenhouse gases greatly affect the temperature of the Earth; without them, Earth's surface would be on average about 33°C (59°F) colder than at present.

"Human activities since the start of the industrial era around 1750 have increased the levels of greenhouse gases in the atmosphere. The 2007 assessment report compiled by the IPCC observed that "changes in atmospheric concentrations of greenhouse gases and aerosols, land cover and solar radiation alter the energy balance of the climate system", and concluded that "increases in anthropogenic greenhouse gas concentrations is very likely to have caused most of the increases in global average temperatures since the mid-20th century."[13]

Are greenhouse gases causing global warming? In my opinion, the earth is a self healer and will survive whatever man or nature throws at it.

For example, if a forest burns it will re-grow and be stronger than before the fire. The Earth

---

[13]   http://en.wikipedia.org/wiki/Greenhouse_gas

has recovered nicely from the manmade disasters of the past and will continue to do so in the future. So, if greenhouse gases are causing or partly causing Global Warming, then in my opinion the Earth will heal itself without human intervention. But never the less, we all should do our own research into Greenhouse Gases and determine within ourselves if global warming is manmade or man made up.

CHAPTER 9
Do We All Agree?

Do Americans agree about global warming? Do you always believe everything that you see in the news? Would it surprise you to know that usually 30 -50 percent of Americans are skeptical regarding facts relayed in a typical news cast. Similar results are found when Americans are asked about Global Warming.

In a recent survey conducted by Yale and George Mason

Universities, 2000 Americans were asked about Global Warming. Here is a summary of the results:

- 18 percent of those surveyed strongly support a variety of climate change policies, such as regulating CO_2 as a pollutant.

- 33 percent aren't taking any steps at all to reduce energy use.

- 7 percent say they're positive that global warming isn't man made and don't worry about the problem.

- 11 percent said they weren't sure global warming was real, but even if it were, it won't happen for a century.

- 12 percent don't give global warming much thought and aren't sure if it's real, or that it poses a threat to future generations.

- More than 50 percent listen to friends and family and don't trust scientists and television weather reporters.

- 19 percent think global warming may be real, but they change their minds. They tend not to think much about issues and believe if anything happens, it won't occur for at least 35 years.

According to a May 2009 article published by *The Christian Science Monitor,* most Americans believe that global warming exists, but do not believe it to be manmade.[14]

From personal experience, I know that this is a controversial issue and opinions are divided. Most of my acquaintances typically trust and agree with the opinions expressed by friends and family members. This normally is a wise choice, but I truly believe that the issue of Global Warming has such a wide future impact on our lives and economy, that it is the responsibility of everyone to become educated and take a firm stand. This is an important issue that needs to transcend partisan politics. It is too important to our future existence to be ignored or left to chance.

Throughout history, seemingly popular decisions and viewpoints have been adopted without due diligence being

---

14 http://features.csmonitor.com/environment/2009/05/20/americans-dont-

performed. These decisions have almost universally ended with disastrous consequences. Decisions made on Global Warming will ultimately impact every person on the planet, all 6 BILLION plus. Do you want to switch to battery powered vehicles, install solar panels or place windmills in your backyard? This is a possibility. It is my contention that all of the facts are not in and that we should not make rash, possibly erroneous decisions. Each and every one of us has to decide if Global

Warming is manmade or man made-up.

Let's take a look at the economic impact. How does Global Warming affect the economy? Emission controls for power plants and automobiles cost millions of dollars for research and development and additional hundreds of millions to produce, install, and maintain the components. Each piece of government legislation places an additional burden of literally hundreds of millions of dollars on the economy. What is the result? An overall reduction in the workforce and higher costs for goods, utilities and transportation.

An additional question that you need to answer is how much inconvenience you are willing to assume? The American consumer will ultimately pay the price in more ways than one. Have you ever considered what it will be like to drive an electric car? In my opinion, these vehicles are very impractical. I don't want a vehicle that needs to be recharged after 200 highway miles. That is just not practical. Nevertheless, this would be the final result.

Vehicles would become impractical to drive and more expensive to operate.

Another often overlooked problem is the impact on the agricultural industry. Cattle produce a large amount of methane. I recently read an article on the front page of USA Today concerning the methane emissions of farm animals and agriculture. The article actually stated that Australia is considering curtailing their beef industry in favor of harvesting kangaroos and feral camels. Why?

Those animals produce less methane gas. Domestically, such restrictions will result in the import of more and more of our foodstuffs because American farmers will not be able to maintain cost effective operations against foreign competition due to increased expenses for environmental concerns.

So I ask you, do you want to import everything? How safe will your food be? What will happen to our national security if we become dependent to foreign nations for more and more of our food supply?

The agricultural industry has been around for thousands of years and no one will convince me that cattle are an actual cause of Global Warming. This is why I am so passionate about this topic. Do you want tofu, soy burgers, and soy milk to become replacements for the real thing? Countries like China and Mexico are producing agricultural products without such constraints. Do they have low methane cows? Are Angus and Guernsey's outdated models? The methane produced in those countries ultimately mixes with our air. We do not live in a vacuum. We are only harming our own economy. The greater our dependency on foreign nations for basic necessities, the more vulnerable we become. If you doubt this, take a look at the petroleum industry. So I ask you once again, is global warming man made or man made up?

CHAPTER 10
Conclusion

In the preceding chapters, I discussed many factors that are contributing to the Global Warming phenomenon. I discussed the possible manipulation of facts and data by individuals guided by self interest and financial gain and I discussed the importance of due diligence when evaluating what one reads or hears in the media. Upon reflection, I have come to my own conclusions. Global Warming does exist and so does Global Cooling.

But is Global Warming man made? This is a much more difficult question to answer. I believe that it is the responsibility of each individual to evaluate the evidence and form their own informed opinion. This is paramount when one considers that changes are being made to the industrial and environmental sectors that will impact each of us financially. Additionally, this debate will have far reaching impact on future generations.

The decisions being made today will affect the children of tomorrow. Is it man made, or is it a naturally occurring pheno-menon?

This is the question that I would like to answer. Many of the topics that I have discussed are usually glanced over or ignored entirely when the topic of Global Warming is discussed in the media mainstream.

Past temperature fluctua-tions on the Earth are also all but ignored. Our technology and machines did not cause global climatic change 20 million years ago when the dinosaurs reigned supreme. Coal burning power plants did not cause the end of the last ice age. I have conducted extensive personal research concerning this topic.

Based on my findings, I have concluded that Global Warming does indeed exist, however, I also believe that what we are seeing today is a naturally occurring climatic change. The Earth is not static, it evolves and constantly changes.

In closing, I recommend that each of you review the data available concerning this topic and formulate your opinion. Only then can you decide if Global Warming is MAN MADE or MAN MADE UP.

APPENDIX A
Additional Resources

The following information is provided as a resource section. It was collected at Wikipedia[15] and is included for your educational benefit in this text.

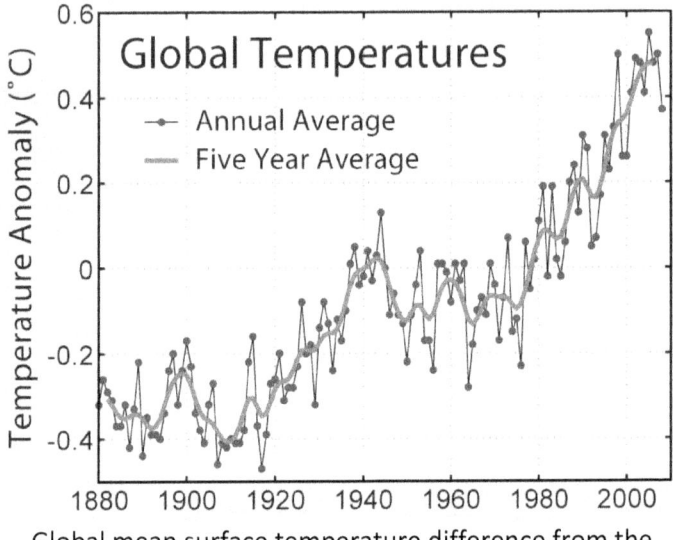

Global mean surface temperature difference from the
average for 1961–1990

1999-2008 Mean Temperatures

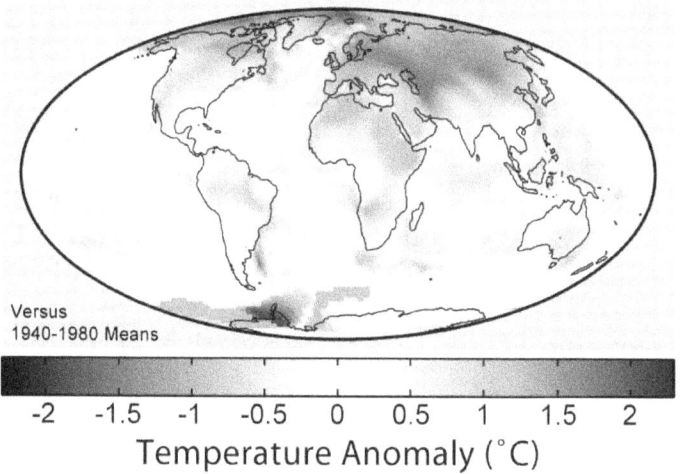

Temperature Anomaly (°C)

Mean surface temperature change for the period 1999 to 2008 relative to the average temperatures from 1940 to 1980

Global warming is the increase in the average temperature of the Earth's near-surface air and oceans since the mid-20th century and its projected continuation. Global surface temperature increased 0.74 ± 0.18°C (1.33 ± 0.32°F) during the last century. The Intergovernmental Panel on Climate Change (IPCC) concludes that increasing greenhouse gas concentrations resulting from human activity such as fossil fuel burning and deforestation caused most of the observed temperature increase since the middle of the

20th century. The IPCC also concludes that variations in natural phenomena such as solar radiation and volcanoes produced most of the warming from pre-industrial times to 1950 and had a small cooling effect afterward. These basic conclusions have been endorsed by more than 45 scientific societies and academies of science, including all of the national academies of science of the major industrialized countries. A small number of scientists dispute the consensus view.

Climate model projections summarized in the latest IPCC report indicate that the global surface temperature will probably rise a further 1.1 to 6.4°C (2.0 to 11.5°F) during the twenty-first century. The uncertainty in this estimate arises from the use of models with differing sensitivity to greenhouse gas concentrations and the use of differing estimates of future greenhouse gas emissions. Some other uncertainties include how warming and related changes will vary from region to region around the globe. Most studies focus on the period up to the year 2100. However, warming is expected to continue beyond 2100 even if emissions stop, because of the large heat capacity of the oceans and the long lifetime of carbon dioxide in the atmosphere.

An increase in global temperature will cause sea levels to rise and will change the amount and pattern of precipitation, probably including expansion of subtropical deserts. The continuing retreat of glaciers, permafrost and sea ice is expected, with warming being strongest in the Arctic. Other likely effects include increases in the intensity of extreme weather events, species extinctions, and changes in agricultural yields.

Political and public debate continues regarding climate change, and what actions (if any) to take in response. The available options are mitigation to reduce further emissions; adaptation to reduce the damage caused by warming; and, more speculatively, geo-engineering to reverse global warming. Most national governments have signed and ratified the Kyoto Protocol aimed at reducing greenhouse gas emissions.

Temperature Changes

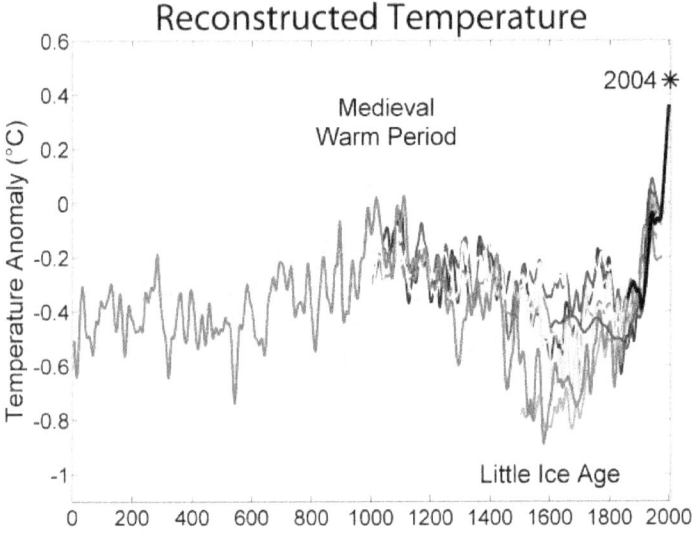

Two millennia of mean surface temperatures according to different reconstructions, each smoothed on a decadal scale. The unsmoothed, annual value for 2004 is also plotted for reference.

The most commonly discussed measure of global warming is the trend in globally averaged temperature near the Earth's surface. Expressed as a linear trend, this temperature rose by 0.74°C ±0.18°C over the period 1906-2005. The rate of warming over the last 50 years of that period was almost double that for the period as a whole (0.13°C ±0.03°C per decade, versus 0.07°C ± 0.02°C per decade). The urban heat island effect is estimated to account for about 0.002°C of warming per decade since 1900. Temperatures in the lower troposphere have increased between 0.12 and 0.22 °C (0.22 and 0.4°F) per decade since 1979, according to satellite temperature measurements. Temperature is believed to have been relatively stable over the one or two thousand years before 1850, with regionally-varying fluctuations such as the Medieval Warm Period or the Little Ice Age.

Based on estimates by NASA's Goddard Institute for Space Studies, 2005 was the warmest year since reliable, widespread instrumental measurements became available in the late 1800s, exceeding the previous record set in 1998 by a few hundredths of a degree. Estimates prepared by the World

Meteorological Organization and the Climatic Research Unit concluded that 2005 was the second warmest year, behind 1998. Temperatures in 1998 were unusually warm because the strongest El Niño in the past century occurred during that year.

Temperature changes vary over the globe. Since 1979, land temperatures have increased about twice as fast as ocean temperatures (0.25°C per decade against 0.13 °C per decade). Ocean temperatures increase more slowly than land temperatures because of the larger effective heat capacity of the oceans and because the ocean loses more heat by evaporation. The Northern Hemisphere warms faster than the Southern Hemisphere because it has more land and because it has extensive areas of seasonal snow and sea-ice cover subject to the ice-albedo feedback. Although more greenhouse gases are emitted in the Northern than Southern Hemisphere this does not contribute to the difference in warming because the major greenhouse gases persist long enough to mix between hemispheres.

The thermal inertia of the oceans and slow responses of other indirect effects mean that climate can take centuries or longer to adjust to changes in forcing. Climate commitment studies indicate that even if greenhouse gases were stabilized at 2000 levels, a further warming of about 0.5 °C (0.9 °F) would still occur.

Radiative Forcing

External forcing is a term used in climate science for processes external to the climate system (though not necessarily external to Earth). Climate responds to several types of external forcing, such as changes in greenhouse gas concentrations, changes in solar luminosity, volcanic eruptions, and variations in Earth's orbit around the Sun. Attribution of recent climate change focuses on the first three types of forcing. Orbital cycles vary slowly over tens of thousands of years and thus are too gradual to have caused the temperature changes observed in the past century.

Greenhouse Gases

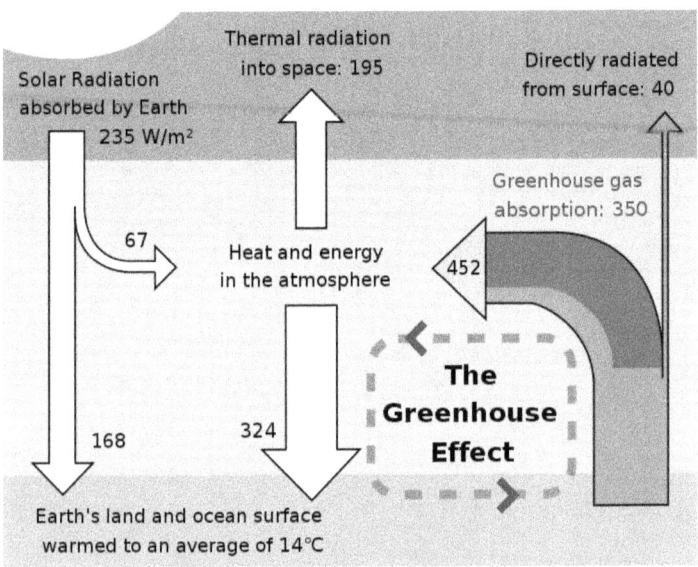

Greenhouse effect schematic showing energy flows between the atmosphere, space, and earth's surface. Energy exchanges are expressed in watts per square meter (W/m^2).

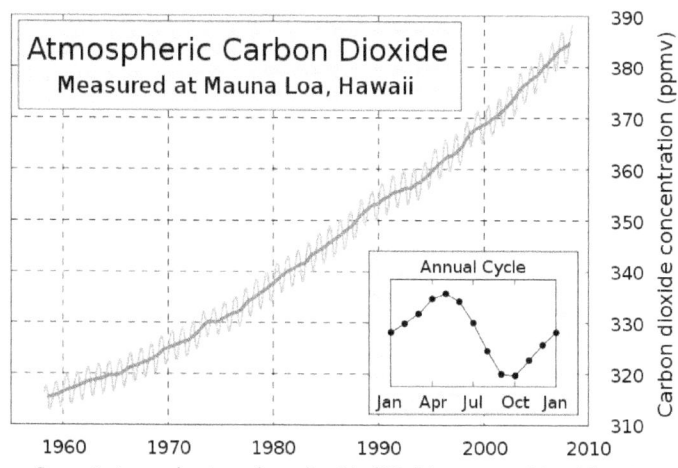

Recent atmospheric carbon dioxide (CO_2) increases. Monthly CO_2 measurements display seasonal oscillations in overall yearly uptrend; each year's maximum occurs during the Northern Hemisphere's late spring, and declines during its growing season as plants remove some atmospheric CO_2.

The greenhouse effect is the process by which absorption and emission of infrared radiation by gases in the atmosphere warm a planet's lower atmosphere and surface. It was discovered by Joseph Fourier in 1824 and was first investigated quantitatively by Svante Arrhenius in 1896. Existence of the greenhouse effect as such is not disputed, even by those who do not agree that the recent temperature increase is attributable to human activity. The question is instead how the strength of the greenhouse effect changes when human activity increases the concentrations of greenhouse gases in the atmosphere.

Naturally occurring green-house gases have a mean warming effect of about 33 °C

(59 °F). The major greenhouse gases are water vapor, which causes about 36–70 percent of the greenhouse effect; carbon dioxide (CO_2), which causes 9–26 percent; methane (CH_4), which causes 4–9 percent; and ozone (O_3), which causes 3–7 percent. Clouds also affect the radiation balance, but they are composed of liquid water or ice and so are considered separately from water vapor and other gases.

Human activity since the Industrial Revolution has increased the amount of greenhouse gases in the atmosphere, leading to increased radiative forcing from CO_2, methane, tropospheric ozone, CFCs and nitrous oxide. The concentrations of CO_2 and methane have increased by 36% and 148% respectively since the mid-1700s. These levels are considerably higher than at any time during the last 650,000 years, the period for which reliable data has been extracted from ice cores. Less direct geological evidence indicates that CO_2 values this high were last seen approximately 20 million years ago. Fossil fuel burning has produced about three-quarters of the increase in CO_2 from human activity over the past 20 years. Most of the rest is due to land-use change, particularly deforestation.

CO_2 concentrations are continuing to rise due to burning of fossil fuels and land-use change. The future rate of rise will depend on uncertain economic, sociological, technological, and natural developments.

Accordingly, the IPCC Special Report on Emissions Scenarios gives a wide range of future CO_2 scenarios, ranging from 541 to 970 ppm by the year 2100. Fossil fuel reserves are sufficient to reach these levels and continue emissions past 2100 if coal, tar sands or methane clathrates are extensively exploited.

The destruction of stratospheric ozone by chlorofluorocarbons is sometimes mentioned in relation to global warming. Although there are a few areas of linkage, the relationship between the two is not strong. Reduction of stratospheric ozone has a cooling influence, but substantial ozone depletion did not occur until the late 1970s. Tropospheric ozone contributes to surface warming.

Aerosols and Soot

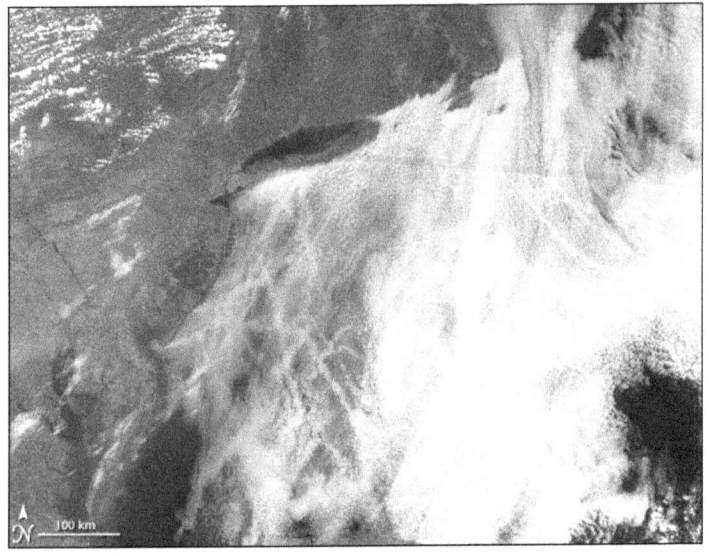

Ship tracks over the Atlantic Ocean on the east coast of the United States. The climatic impacts from aerosol forcing could have a large effect on climate through the indirect effect.

Global dimming, a gradual reduction in the amount of global direct irradiance at the Earth's surface, has partially counteracted global warming from 1960 to the present. The main cause of this dimming is aerosols produced by volcanoes and pollutants. These aerosols exert a cooling effect by increasing the reflection of incoming sunlight. James Hansen and colleagues have proposed that the effects of the products of fossil fuel combustion—CO_2 and aerosols—have largely offset one another in recent decades, so that net warming has been driven mainly by non-CO_2 greenhouse gases.

In addition to their direct effect by scattering and absorbing solar radiation, aerosols have indirect effects on the radiation budget. Sulfate aerosols act as cloud condensation nuclei and thus lead to clouds that have more and smaller cloud droplets. These clouds reflect solar radiation more efficiently than clouds with fewer and larger droplets. This effect also causes droplets to be of more uniform size, which reduces growth of raindrops by collision-coalescence. Clouds modified by pollution have been shown to produce less drizzle, making the cloud brighter and more reflective to incoming sunlight, especially in the near-infrared part of the spectrum.

Soot may cool or warm, depending on whether it is airborne or deposited. Atmospheric soot aerosols directly absorb solar radiation, which heats the atmosphere and cools the surface. Regionally (but not globally), as much as 50% of surface warming due to greenhouse gases may be masked by atmospheric brown clouds. When deposited, espccially on glaciers or on ice in arctic regions, the lower surface albedo can also directly heat the surface. The influences of aerosols, including black carbon, are most pronounced in the tropics and sub-tropics, particularly in Asia, while the effects of greenhouse gases are dominant in the extratropics and southern hemisphere.

Solar Variation

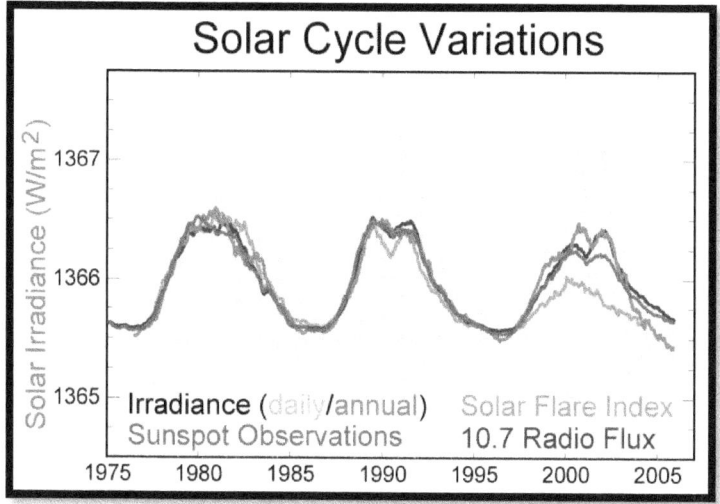

Solar variation over the last thirty years.

Variations in solar output have been the cause of past climate changes. Although solar forcing is generally thought to be too small to account for a significant part of global warming in recent decades, a few studies disagree, such as a recent phenomenological analysis that indicates the contribution of solar forcing may be underestimated.

Greenhouse gases and solar forcing affect temperatures in different ways. While both increased solar activity and increased greenhouse gases are expected to warm the troposphere, an increase in solar activity should warm the stratosphere while an increase in greenhouse gases should cool the stratosphere. Observations show that temperatures in the stratosphere have been steady or cooling since 1979, when satellite measurements became available. Radiosonde (weather balloon) data from the pre-satellite era show cooling since 1958, though there is greater uncertainty in the early radiosonde record.

A related hypothesis, proposed by Henrik Svensmark, is that magnetic activity of the sun deflects cosmic rays that may influence the generation of cloud condensation nuclei and thereby affect the climate. Other research has found no relation between warming in recent decades and cosmic rays. A recent study concluded that the influence of cosmic rays on cloud cover is about a factor of 100 lower than needed to explain the observed changes in clouds or to be a significant contributor to present-day climate change.

Feedback

A positive feedback is a process that amplifies some change. Thus, when a warming trend results in effects that induce further warming, the result is a positive feedback; when the warming results in effects that reduce the original warming, the result is a negative feedback. The main positive feedback in global warming involves the tendency of warming to increase the amount of water vapor in the atmosphere. The main negative feedback in global warming is the effect of temperature on emission of infrared radiation: as the temperature of a body increases, the emitted radiation increases with the fourth power of its absolute temperature.

Water vapor feedback

If the atmosphere is warmed, the saturation vapor pressure increases, and the amount of water vapor in the atmosphere will tend to increase. Since water vapor is a greenhouse gas, the increase in water vapor content makes the atmosphere warm further; this warming causes the atmosphere to hold still more water vapor (a positive feedback), and so on until other processes stop the feedback loop. The result is a much larger greenhouse effect than that due to CO_2 alone.

Although this feedback process causes an increase in the absolute moisture content of the air, the relative humidity stays nearly constant or even decreases slightly because the air is warmer.

Cloud feedback

Warming is expected to change the distribution and type of clouds. Seen from below, clouds emit infrared radiation back to the surface, and so exert a warming effect; seen from above, clouds reflect sunlight and emit infrared radiation to space, and so exert a cooling effect. Whether the net effect is warming or cooling depends on details such as the type and altitude of the cloud, details that are difficult to represent in climate models.

Lapse rate

The atmosphere's temperature decreases with height in the troposphere. Since emission of infrared radiation varies with the fourth power of temperature, longwave radiation escaping to space from the relatively cold upper atmosphere is less than that emitted toward the ground from the lower atmosphere. Thus, the strength of the greenhouse effect depends on the atmosphere's rate of temperature decrease with height. Both theory and climate models indicate that global warming will reduce the rate of temperature decrease with height, producing a negative *lapse rate feedback* that weakens the greenhouse effect. Measurements of the rate of temperature change with height are very sensitive to small errors in observations, making it difficult to establish whether the models agree with observations.

Ice-albedo feedback

Aerial photograph showing a section of sea ice. The lighter areas are melt ponds and the darkest areas are open water, both have a lower albedo than the white sea ice. The melting ice contributes to the ice-albedo feedback.

When ice melts, land or open water takes its place. Both land and open water are on average less reflective than ice and thus absorb more solar radiation. This causes more warming, which in turn causes more melting, and this cycle continues.

Arctic methane release

Warming is also the triggering variable for the release of methane from sources both on land and on the deep ocean floor, making both of these possible feedback effects. Thawing permafrost, such as the frozen peat bogs in Siberia, creates a positive feedback due to the release of CO_2 and methane.

Reduced absorption of CO2 by the oceans

Ocean ecosystems' ability to sequester carbon is expected

to decline as the oceans warm. This is because warming reduces the nutrient levels of the mesopelagic zone (about 200 to 1000 m deep), which limits the growth of diatoms in favor of smaller phytoplankton that are poorer biological pumps of carbon.

Gas release

Release of miscellaneous gases of biological origin may be affected by global warming, but research into such effects is at an early stage. Such releases may have direct climate effects, such as

Nitrous oxide released from peat and indirect effects, such as Dimethyl sulfide released from oceans.

Climate models

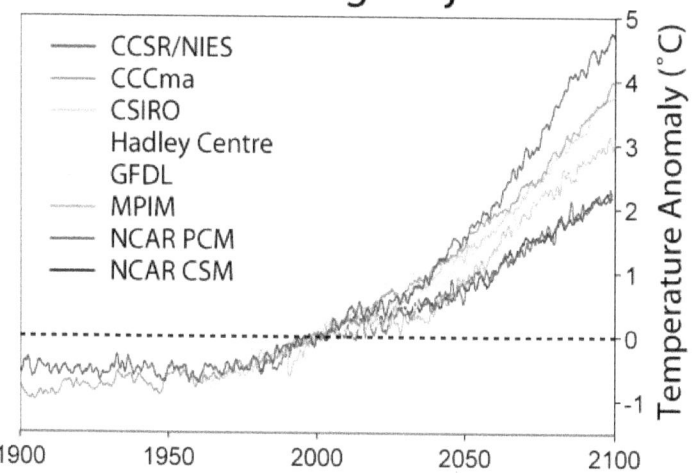

Calculations of global warming prepared in or before 2001 from a range of climate models under the SRES A2 emissions scenario, which assumes no action is taken to reduce emissions and regionally divided economic development.

Global Warming Predictions

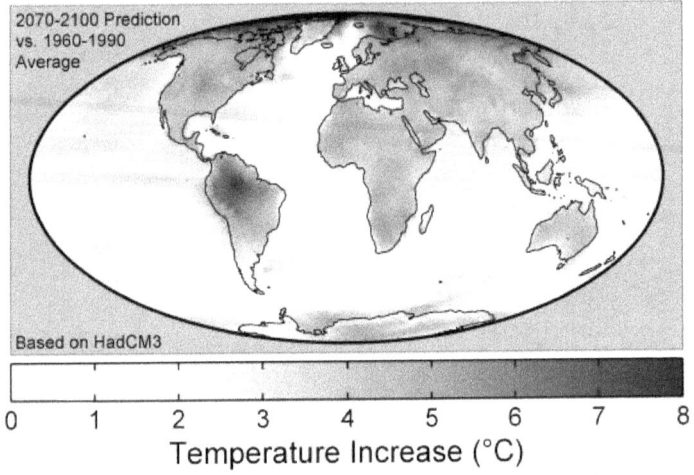

2070-2100 Prediction
vs. 1960-1990
Average

Based on HadCM3

0   1   2   3   4   5   6   7   8

Temperature Increase (°C)

The geographic distribution of surface warming during the 21st century
calculated by the HadCM3 climate model if a business as usual scenario is
assumed for economic growth and greenhouse gas emissions. In this
figure, the globally averaged warming corresponds to 3.0°C (5.4°F).

The main tools for projecting
future climate changes are
computer models of the climate.
These models are based on
physical principles including fluid
dynamics and radiative transfer.
Although they attempt to include

140

as many processes as possible, simplifications of the actual climate system are inevitable because of the constraints of available computer power and limitations in knowledge of the climate system. All modern climate models are in fact *combinations* of models for different parts of the Earth that are coupled to one another. These include an atmospheric model for air movement, temperature, clouds, and other atmospheric properties; an ocean model that predicts temperature, salt content, and circulation of ocean waters; models for ice cover on land and sea; and a model of heat and moisture transfer from soil and vegetation to the atmosphere. Some models also include treatments of chemical and biological processes. Climate models project a warmer climate due to increasing levels of greenhouse gases. Although much of the variation in model outcomes depends on the greenhouse gas emissions used as inputs, the temperature effect of a specific greenhouse gas concentration (climate sensitivity) varies depending on the model used. The representation of clouds is one of the main sources of uncertainty in present-generation models.

Global climate model projections of future climate most often have used estimates of greenhouse gas emissions from the IPCC Special Report on Emissions Scenarios (SRES). In addition to human-caused emissions, some models also include a simulation of the carbon cycle; this generally shows a positive feedback, though this response is uncertain. Some observational studies also show a positive feedback.

Including uncertainties in future greenhouse gas concentrations and climate sensitivity, the IPCC anticipates a warming of 1.1°C to 6.4°C (2.0°F to 11.5°F) by the end of the 21st century, relative to 1980–1999. A 2008 paper predicts that the global temperature may not increase during the next decade because short-term natural fluctuations may temporarily outweigh greenhouse gas-induced warming.

Models are also used to help investigate the causes of recent climate change by comparing the observed changes to those that the models project from various natural and human-derived causes. Although these models do not unambiguously attribute the warming that occurred from approximately 1910 to 1945 to either natural variation or human effects, they do indicate that the warming since 1975 is dominated by man-made greenhouse gas emissions.

Current climate models produce a good match to observations of global temperature changes over the last century, but do not simulate all aspects of climate. The physical realism of models is tested by examining their ability to simulate current or past climates. While a 2007 study by David Douglass and colleagues found that the models did not accurately predict observed changes in the tropical troposphere, a 2008 paper published by a 17-member team led by Ben Santer noted errors in the Douglass study, and found instead that the models and observations were not statistically different. Not all effects of global warming are accurately predicted by the climate models used by the IPCC. For example, observed Arctic shrinkage has been faster than that predicted.

Attributed and Expected Effects

Environmental

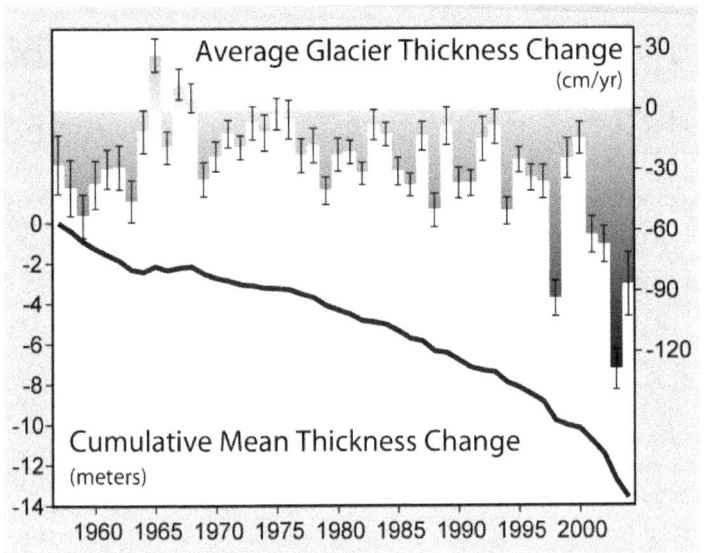

Sparse records indicate that glaciers have been retreating since the early 1800s. In the 1950s measurements began that allow the monitoring of glacial mass balance, reported to the WGMS and the NSIDC.

It usually is impossible to connect specific weather events to global warming. Instead, global warming is expected to cause changes in the overall distribution and intensity of events, such as changes to the frequency and intensity of heavy precipitation. Broader effects are expected to include glacial retreat, Arctic shrinkage.

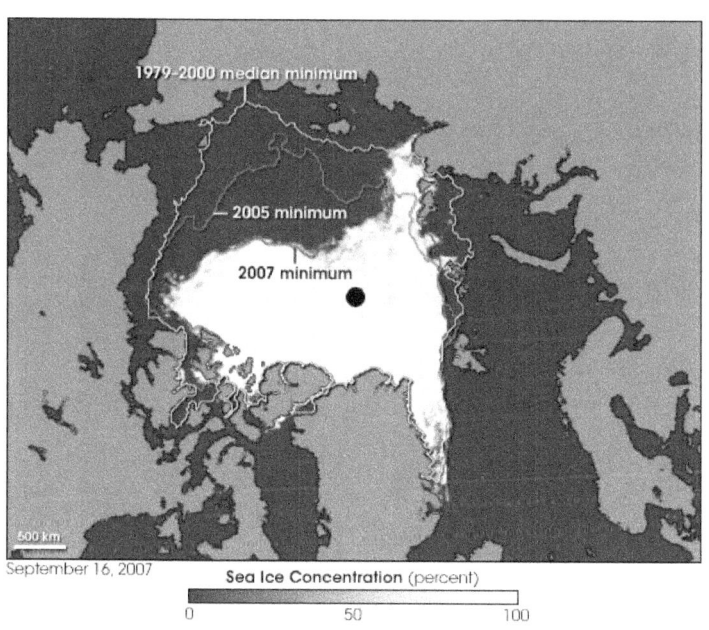

To order additional copies of
this book, visit
amazon.com.

*A certain percentage of the
sales of this book will go
towards environmental
awareness issues and hospice
organizations.*

www.ingramcontent.com/pod-product-compliance
Lightning Source LLC
Chambersburg PA
CBHW051529170526

45165CB00002B/662